DIG DEEP INTO FOSSILS

Digging Up Petrified Forests

BY CHARLOTTE TAYLOR

Enslow PUBLISHING

INVESTIGATE!

Please visit our website, www.enslow.com. For a free color catalog of all our high-quality books, call toll free 1-800-398-2504 or fax 1-877-980-4454.

Library of Congress Cataloging-in-Publication Data

Names: Taylor, Charlotte, 1978- author.
Title: Digging up petrified forests / Charlotte Taylor.
Description: New York : Enslow Publishing, 2022. | Series: Dig deep into fossils | Includes index.
Identifiers: LCCN 2020030844 (print) | LCCN 2020030845 (ebook) | ISBN 9781978521643 (library binding) | ISBN 9781978521629 (paperback) | ISBN 9781978521636 (set) | ISBN 9781978521650 (ebook)
Subjects: LCSH: Petrified forests–Juvenile literature.
Classification: LCC QE991 .T39 2022 (print) | LCC QE991 (ebook) | DDC 561/.16-dc23
LC record available at https://lccn.loc.gov/2020030844
LC ebook record available at https://lccn.loc.gov/2020030845

Published in 2022 by
Enslow Publishing
101 West 23rd Street, Suite #240
New York, NY 10011

Designer: Katelyn E. Reynolds
Editor: Megan Quick

Photo credits: Cover, pp. 1 (petrified log) Calesh/Shutterstock.com; cover, pp. 1–24 (background) Click Bestsellers/Shutterstock.com; cover, pp. 1–24 (tools) Serz_72/ Shutterstock.com; cover, pp. 1–24 (burst) Foxys Graphic/Shutterstock.com; p. 4 MyLoupe/Universal Images Group via Getty Images; p. 5 KiraVolkov/ iStock / Getty Images Plus; p. 5 alptraum/ iStock / Getty Images Plus; p. 5 SweetyMommy/ iStock / Getty Images Plus; p. 5 milehightraveler/ iStock / Getty Images Plus; p. 6 Brian van der Brug/Los Angeles Times via Getty Images; p. 7 miralex/ iStock / Getty Images Plus; p. 8 Tom Pfeiffer / VolcanoDiscovery/Moment/Getty Images; p. 9 WeiliLi/ iStock / Getty Images Plus; p. 11 (water) anucha sirivisansuwan/Moment/Getty Images; p. 11 (sand) the_burtons/Moment/Getty Images; p. 11 (ash) Roman Stavila/ iStock / Getty Images Plus; p. 11 (mud) JKunnen/ iStock / Getty Images Plus; p. 11 (log) Barcin/E+/Getty Images; p. 12 Bjorn Holland/ The Image Bank / Getty Images Plus; p. 13 Mark Boster/Los Angeles Times via Getty Images; pp. 14, 15 (quartz and iron oxide), 21 Arterra/Universal Images Group via Getty Images; p. 15 (manganese oxide) miriam-doerr/ iStock / Getty Images Plus; p. 16 skibreck/ iStock / Getty Images Plus; p. 17 DeAgostini/Getty Images; p. 18 David Clapp/ Stone/Getty Images; p. 19 Alinabel/ iStock / Getty Images Plus; p. 20 Joe McDonald/ The Image Bank/Getty Images; p. 22 Hiroyuki Matsumoto/ Photolibrary / Getty Images Plus; p. 23 Witold Skrypczak/ Lonely Planet Images / Getty Images Plus; p. 24 Danita Delimont/ Gallo Images / Getty Images Plus; p. 25 David H. Carriere/ The Image Bank / Getty Images Plus; p. 25 Melissa Kopka/ iStock / Getty Images Plus; p. 26 Ruben Earth/Moment/Getty Images; p. 27 Holger Leue/ The Image Bank Unreleased/Getty Images; p. 28 Lyn Alweis/The Denver Post via Getty Images; p. 29 Frank Staub/ Photolibrary / Getty Images Plus.

Portions of this work were originally authored by Kathleen Connors and published as *Petrified Forests*. All new material in this edition authored by Charlotte Taylor.

Printed in the United States of America

CPSIA compliance information: Batch #CSENS22: For further information contact Enslow Publishing, New York, New York, at 1-800-398-2504.

CONTENTS

WORDS IN THE GLOSSARY APPEAR IN **BOLD** TYPE THE FIRST TIME THEY ARE USED IN THE TEXT.

A DIFFERENT KIND OF FOREST

When you think of forests, you probably think of tall, green trees made of wood. But there's another kind of forest. A petrified forest is made of ancient trees whose wood has turned to stone. Many trees in petrified forests are millions of years old.

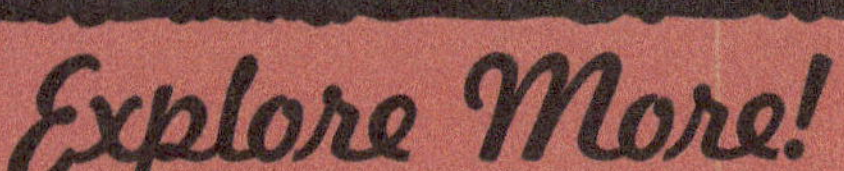

PETRIFIED FORESTS CONTAIN FOSSILS OF ANIMALS AND OTHER PLANTS AS WELL AS TREES.

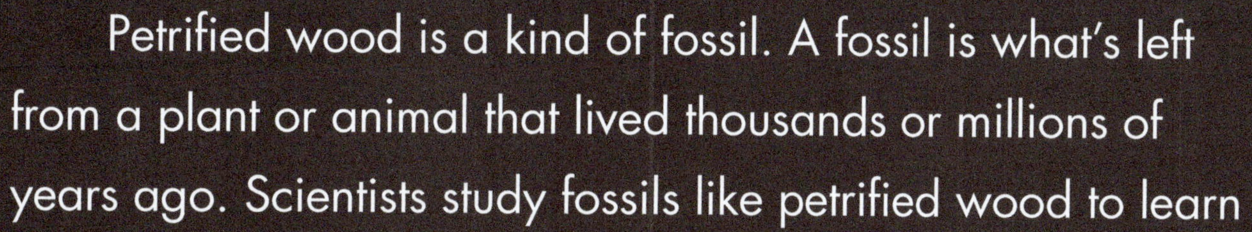

Petrified wood is a kind of fossil. A fossil is what's left from a plant or animal that lived thousands or millions of years ago. Scientists study fossils like petrified wood to learn what Earth was like long ago and how it has changed.

PIECES OF WOOD IN A PETRIFIED FOREST LIKE THIS ONE CAN BE OVER 200 MILLION YEARS OLD!

FROM TREE TO STONE

Most of the time, when a tree dies, it simply rots. As it falls apart, it becomes part of the soil. So how does a tree turn into stone? A few things must happen.

THIS FOSSILIZED TREE WAS FOUND BURIED IN ROCK IN UTAH. THE ROCK STARTED AS SEDIMENT AND HARDENED OVER TIME.

First, the tree must be buried soon after it dies. If it is not buried quickly, then it will start to break down. Trees may be buried in sediment—bits of rock, sand, and other matter—that's carried by water. A lot of petrified wood is found near rivers and other bodies of water. The sediment protects the tree.

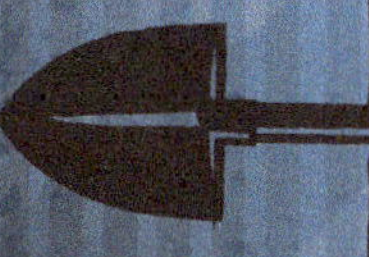

Explore More!

FOSSILIZED WOOD CAN BE FOUND THROUGHOUT THE UNITED STATES. TODAY, MANY PETRIFIED FORESTS ARE NATIONAL PARKS.

Sediment can also come from a **volcano**. When a volcano **erupts**, everything around it is covered with a blanket of ash and mud. This **layer** of sediment **preserves** the trees.

A VOLCANO MAY COVER MILES OF GROUND WITH SEDIMENT WHEN IT ERUPTS.

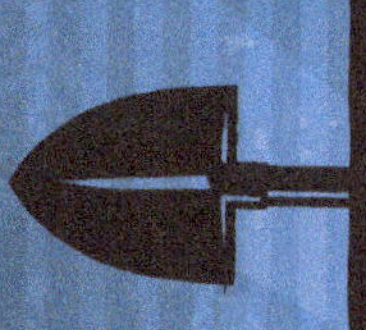

Explore More!

IT TAKES MILLIONS OF YEARS FOR WOOD TO PETRIFY IN NATURE, BUT SCIENTISTS HAVE FIGURED OUT HOW TO PETRIFY WOOD IN A FEW DAYS.

THIS PETRIFIED TREE STANDS IN YELLOWSTONE NATIONAL PARK.

The sediment from water or a volcano must bury the tree deep enough that it does not get much **oxygen** from the air. This slows down rotting. Now a fossil can form. Some petrified wood fossils are **casts**. They form when the wood slowly breaks down over a very long time. **Minerals** fill the spaces the wood left and take on its shape.

PERMINERALIZATION

When minerals flow into wood and harden, it's called permineralization (puhr-mih-nuh-ruhl-ih-ZAY-shun). The minerals may come from ash or water in the ground. They fill spaces inside and between **cells** in the wood. After a while, the minerals harden into rock.

Petrified forests in the western United States are made of wood permineralized by a mineral called silica. Millions of years ago, the trees were buried in sediment created by volcanic ash that had silica in it. When it mixed with water in the ground, the silica moved into the wood and started permineralization.

HOW PERMINERALIZATION WORKS

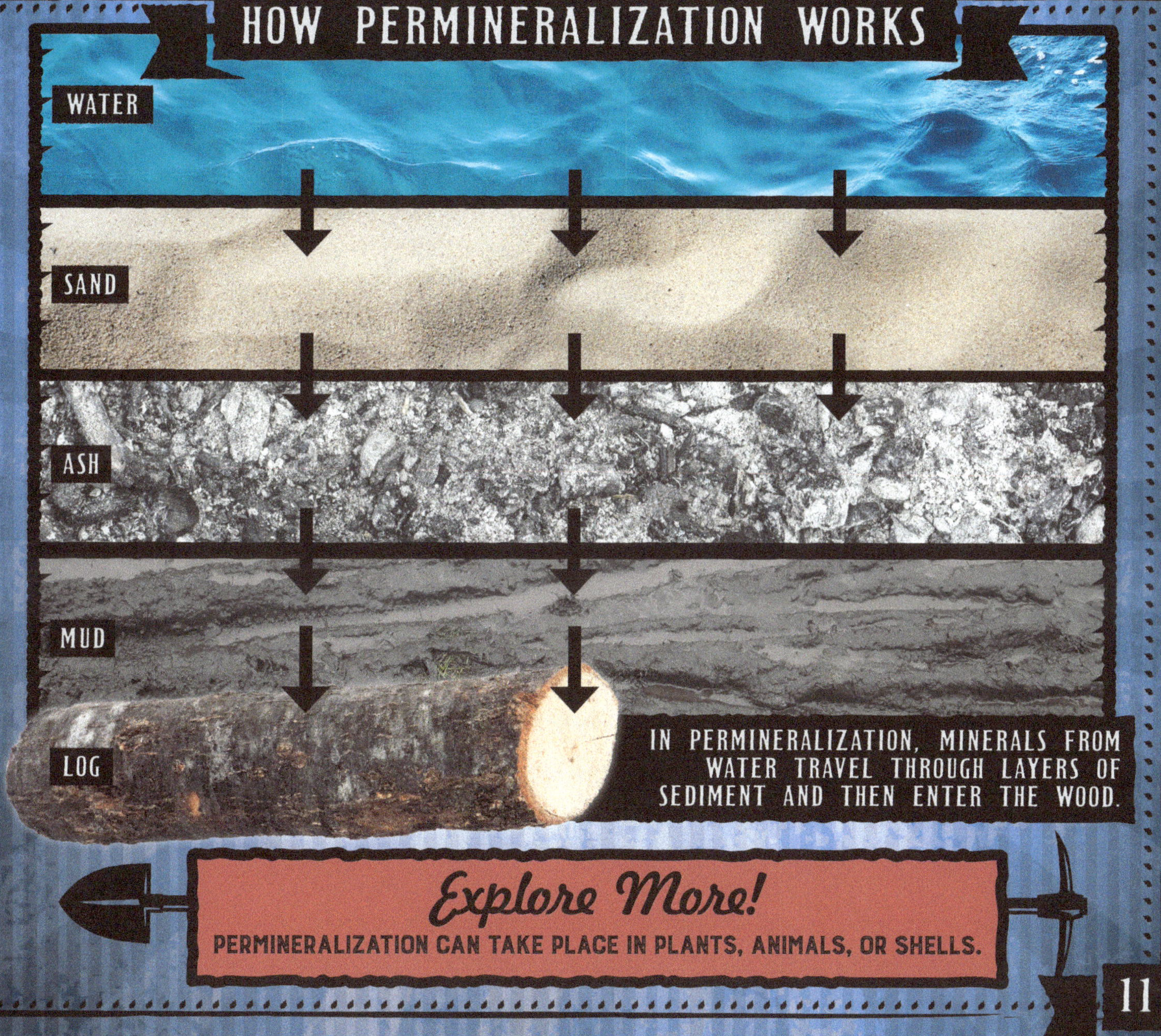

IN PERMINERALIZATION, MINERALS FROM WATER TRAVEL THROUGH LAYERS OF SEDIMENT AND THEN ENTER THE WOOD.

Explore More!

PERMINERALIZATION CAN TAKE PLACE IN PLANTS, ANIMALS, OR SHELLS.

FINDING FOSSILS

Sometimes fossils are not hard to find. Some petrified forests can be seen today because the ground shifted over time and lifted up areas where the trees were buried. The sediment and rock covering them wore away. But many fossils remain buried. Scientists and other fossil hunters must dig them up.

THIS PETRIFIED WOOD IN ARIZONA IS NO LONGER COVERED IN SEDIMENT.

Paleontologists are scientists who study fossils to learn what life was like long ago. They use special tools to dig up fossils. They are careful not to break them. It can take a paleontologist years to fully study a fossil.

KIM SCOTT, A PALEONTOLOGIST AND GEOLOGIST, HOLDS AN ANCIENT PINE BRANCH THAT HAS BEEN DUG UP.

Explore More!

GEOLOGISTS ARE SCIENTISTS WHO LEARN ABOUT EARTH'S HISTORY BY STUDYING ROCKS AND OTHER MATTER. THEY ALSO WORK WITH FOSSILS.

IN FULL COLOR

Some petrified wood is a lot more colorful than an everyday log. The silica in the fossil becomes quartz, which is a mineral that forms crystals. Quartz can change color based on traces of other minerals in it. Fossilized wood may include a rainbow of colors! However, some petrified wood still looks like regular brown logs.

DIFFERENT MINERALS IN THIS PIECE OF PETRIFIED WOOD CAUSE IT TO HAVE SHADES OF BLUE, RED, AND YELLOW.

Explore More!

THERE ARE THREE MAIN MINERALS THAT GIVE PETRIFIED WOOD ITS COLORS: QUARTZ (WHITE), MANGANESE OXIDE (BLUE, PURPLE, BLACK, AND BROWN), AND IRON OXIDE (YELLOW, RED, AND BROWN).

QUARTZ

IRON OXIDE

Petrified forests don't look like modern forests. Most of the trees are lying down and have been broken into pieces. They've lost all their branches, leaves, and bark.

A TREE'S AGE

Many living trees have growth rings. These are the rings that you can see when the tree's trunk is cut open. Each ring stands for one growth period. Older trees will have more growth rings than younger trees.

THE GROWTH RINGS IN THIS PIECE OF PETRIFIED WOOD CAN BE CLEARLY SEEN.

Trees don't grow after they die, so growth rings will not tell the age of a fossilized tree. That does not mean that growth rings aren't useful. The rings can help scientists understand what the tree's **environment** was like in ancient times, like its **climate** and seasons.

Explore More!

THE METHOD OF FIGURING OUT DATES BASED ON TREES' GROWTH RINGS IS CALLED DENDROCHRONOLOGY.

FOSSIL DATING

Scientists have special ways of figuring out how old a fossil is. One method is to look at the rock where the fossil was buried. The lower or deeper a rock layer, the older the rock. The fossil's age can be figured from the age of the rock.

THIS LOG IS PART OF AN ANCIENT PETRIFIED FOREST.

Explore More!

PETRIFIED FORESTS EXIST ALL AROUND THE WORLD. ONE OF THE MOST FAMOUS PETRIFIED FORESTS IS MAADI PETRIFIED FOREST IN EGYPT. THE FOSSILIZED TREES THERE DATE BACK 35 MILLION YEARS!

EARTH'S TIME PERIODS

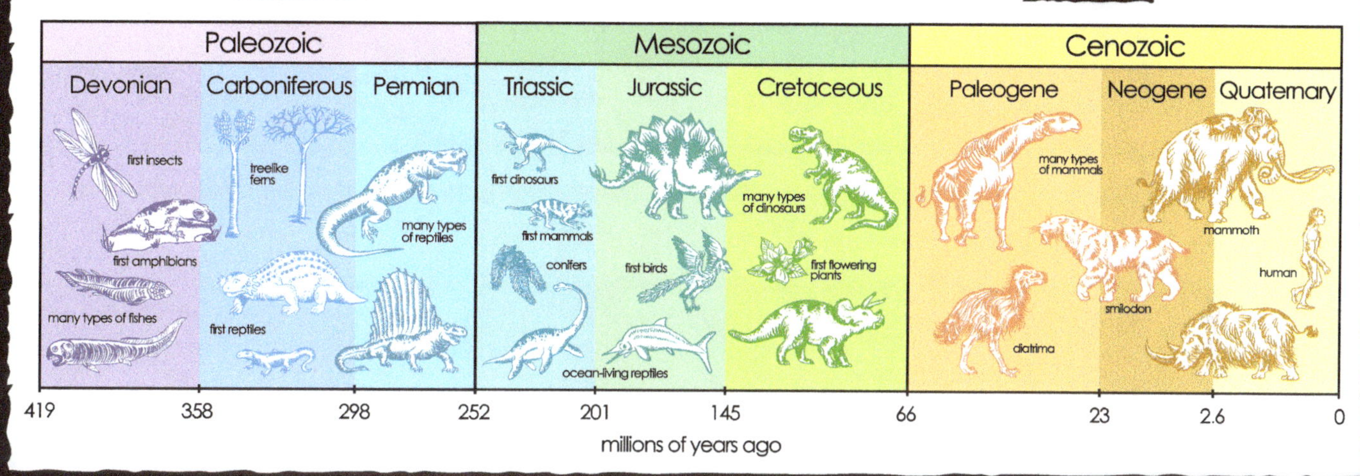

SCIENTISTS BREAK DOWN EARTH'S HISTORY INTO TIME PERIODS. TREES GO BACK ALMOST 400 MILLION YEARS TO THE DEVONIAN **ERA**.

Some petrified wood is super old! It depends on when the forests were buried by sediment or a nearby volcano erupted. For example, a petrified forest in California is about 3.4 million years old, while one in Arizona is more than 211 million years old!

PIECES OF THE PUZZLE

Why are petrified forests important? They can help us piece together the puzzle of what the world used to be like. By comparing cells of fossilized trees to modern trees, scientists can figure out when certain trees began to grow in an area.

PETRIFIED WOOD MAY CONTAIN BITS OF ORIGINAL MATTER THAT IS MILLIONS OF YEARS OLD.

Petrified wood can also be used to figure out what an area's climate was like millions of years ago. The wood, plus other plant and animal fossils found near it, can tell scientists if an area that's now a desert used to be a forest or a riverbed!

THIS PETRIFIED TREE SITS NEAR AN OLD RIVERBED. SCIENTISTS CAN STUDY THE AREA AROUND FOSSILS TO FIGURE OUT MORE OF THEIR HISTORY.

Explore More!

THE PIECES OF WOOD IN PETRIFIED FORESTS OFTEN LOOK LIKE SOMEONE CUT THEM WITH A SAW. BUT THEY ONLY LOOK THAT WAY BECAUSE THE QUARTZ IN THEM IS HARD AND BREAKS VERY CLEANLY.

HOW ONE FOREST FORMED

There are petrified forests all across the United States. One of the largest in the world is Petrified Forest National Park in Arizona. The petrified trees there formed because many trees grew along a river long ago. After the trees died, they floated down the river and got caught together in **logjams**. In time, these logjams became the different "forests" in the park.

IT DOESN'T LOOK LIKE IT NOW, BUT PETRIFIED FOREST NATIONAL PARK IN ARIZONA USED TO BE THE SITE OF A LARGE RIVER SYSTEM.

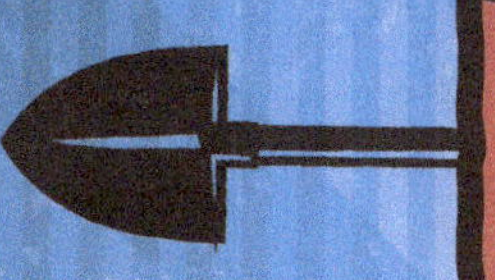

Explore More!

SOME OF THE TREES IN PETRIFIED FOREST NATIONAL PARK LIVED 218 MILLION YEARS AGO!

Long ago, there were volcanoes to the west and south of the forest. Eruptions allowed ash and silica to enter the rivers and build up. This matter found its way into the trees in the river.

A NATIONAL TREASURE

Scientists have **identified** about a dozen kinds of petrified trees in Petrified Forest National Park. They've also found that the trees were pretty huge: Some were at least 200 feet (61 m) tall when they were alive!

MINERALS HAVE CAUSED CRYSTALS TO FORM IN THESE LOGS AT PETRIFIED FOREST NATIONAL PARK.

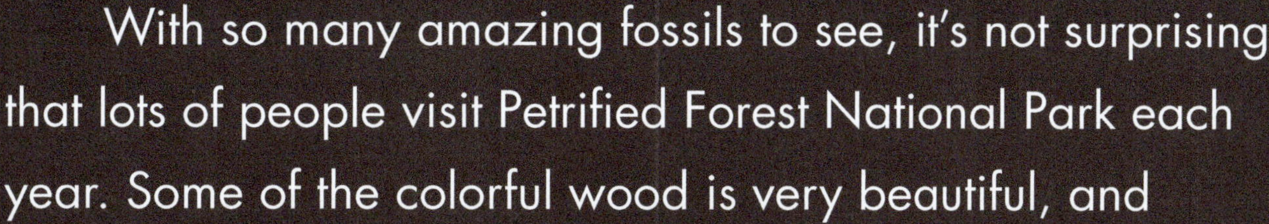

With so many amazing fossils to see, it's not surprising that lots of people visit Petrified Forest National Park each year. Some of the colorful wood is very beautiful, and people want to take pieces of it home with them. But because it's a national park, it's against the law to take a fossil from the park.

Explore More!

TREES, PLANTS, AND ANIMALS AREN'T THE ONLY THINGS IN PETRIFIED FOREST NATIONAL PARK. TRACES OF HUMANS IN THE AREA DATE BACK 10,000 YEARS!

POLAR FORESTS

Sometimes fossil forests appear in unlikely places. In 2016, paleontologists discovered fossilized trees in the icy land of Antarctica. It might be hard to picture now, but millions of years ago, the South Pole was much warmer—and greener! Huge trees that were 65 to 131 feet (20 to 40 m) tall covered the area.

SNOWY ANTARCTICA WAS THE HOME OF LARGE, GREEN FORESTS UNTIL ABOUT 14 MILLION YEARS AGO.

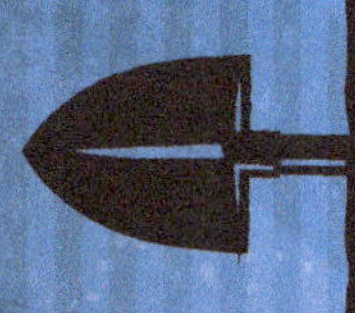

Explore More!

SCIENTISTS HOPE FOSSILS WILL HELP THEM UNDERSTAND HOW ANCIENT TREES IN ANTARCTICA LIVED DURING COLD WINTERS WITH VERY LITTLE SUNLIGHT.

The tree fossils found by the paleontologists are probably about 280 million years old. The scientists believe that volcanoes wiped out the forest and buried it under ash. The layer of ash helped preserve the trees and form fossils.

VISITORS TO ANTARCTICA VIEW ANCIENT FOSSILS.

OLDER THAN THE DINOSAURS

The Gilboa Fossil Forest in New York State is the world's oldest forest. The trees in this amazing forest go back 385 million years! Even though the first tree stumps were found in the 1920s, scientists were not able to clear the area and study the whole forest until 2010.

SCIENTISTS DIG IN SEARCH OF TREE FOSSILS THAT WILL ANSWER MORE QUESTIONS ABOUT THE HISTORY OF TREES.

EACH PIECE OF PETRIFIED WOOD CONTAINS CLUES TO THE PAST.

Petrified forests like those in New York and Arizona are important for what they tell us about the trees themselves as well as the area where they lived. Scientists hope to make even more discoveries as they continue to learn more about the world millions of years ago.

GLOSSARY

cast A fossil that is made when a mark or hole left behind by something is filled by minerals.

cell The smallest unit that makes up a living thing.

climate The average weather conditions of a place over a period of time.

environment The natural world in which a plant or animal lives.

era A period of time that is known for a certain event or quality.

erupt To burst forth.

identify To find out the name or features of something.

layer One thickness of something lying over or under another.

logjam A group of logs stuck together in a waterway.

mineral Matter found in nature that is not living.

oxygen A colorless, odorless gas that many animals, including people, need to breathe.

preserve To keep safe.

volcano An opening in a planet's surface through which hot, liquid rock sometimes flows.

For More Information

BOOKS

Alkire, Jessie. *Discovering Fossils.* Minneapolis, MN: Abdo Publishing, 2019.

Levy, Janey. *20 Fun Facts About Fossils.* New York, NY: Gareth Stevens, 2018.

Oxlade, Chris. *Rocks and Fossils.* New York, NY: Kingfisher, 2018.

WEBSITES

DK Find Out: Fossil Facts for Kids
www.dkfindout.com/us/dinosaurs-and-prehistoric-life/fossils/
Find out more about fossils, where they are found, and how they are formed.

National Park Service: Become a Junior Ranger
www.nps.gov/kids/become-a-junior-ranger.htm
Find out how you can become a junior ranger and help take care of our national parks!

Petrified Forest National Park for Kids
www.nps.gov/pefo/forkids/index.htm
Learn more about the park through photos, games, and activities.

INDEX